SHEMEN
שֶׁמֶן

OILS OF THE ANCIENT WORLD

JOSHUA GRAFF, MAR

Copyright © 2017 by Joshua Graff. Book design copyright © 2017 Growing Healthy Homes LLC™. Editing by Modern Day Wordsmith. All rights reserved. No part of this book may be reproduced or transmitted in any form or by any means, electronic, mechanical, photocopying or recording without the express written permission of the author and publisher. Scripture quotations taken from the New American Standard Bible® (NASB), Copyright © 1960, 1962, 1963, 1968, 1971, 1972, 1973, 1975, 1977, 1995 by The Lockman Foundation. Used by permission. www.Lockman.org

ISBN: 978-0-9988534-1-3
Printed in the United States of America
First printing.

Growing Healthy Homes LLC
P.O. Box 3154
Bartlesville, OK 74006

To obtain additional copies of this book, please visit
www.GrowingHealthyHomes.com.

Disclaimer: The information presented in the book is for educational purposes only. It is not intended to diagnose, treat, cure, or prevent any disease or illness. If you have a medical condition, please consult the health care provider of your choice.

The author is not liable for the misuse or misunderstanding of any information contained within this publication.

After extensive research, the author has concluded that Young Living Essential Oils are 100% pure, authentic, and unadulterated. This is supported by their proprietary Seed to Seal® process. The author encourages you to do your own research and be an informed consumer.

To my beautiful wife Sharon.
I am truly blessed to be your husband.

Table of Contents

Introduction 11
What Is An Essential Oil? 15
    Ways to Use Essential Oils 15
Historical Facts 17
    Ancient Spice Trade Route 20
Usage in the Bible 23
    Anointing 23
    Types of Anointing 23
    Perfume and Incense 26
    Condiment, Ointment, and Wellness 27
Bible Oils 29
    Exact Bible Oils 30
    Equivalent Bible Oils 69
    Blends 84
Conclusion 87
All the Plants of the Bible 89
Jewish Recipes Using Vitality™ Oils 93
Diffuser Blend Recipes 97
Endnotes 99
Bibliography 101
About the Author 105
Index 107

Acknowledgments

First, I must give all the glory to God, my Savior, for without Him, I would be nothing. He is the Creator and Sustainer of my life, and I know that true life change only comes through a relationship with Him.

I must also thank my amazing wife Sharon for all the support she has given through this process, especially putting up with me when I've been ultra-focused on my writing. Sharon, you are a blessing. I love you more every day as we travel the world and change lives together.

Thank you to my parents Billy and Dawn Graff for their constant encouragement to pursue my passions, as well as all they taught me as I was growing up.

Special thanks to my mother-in-law Debra Raybern for her great insights, her help with editing, and for writing a foreword for this book.

Thank you to Adaryll Jordan for writing a second foreword for me.

I must thank all my family members who helped edit, revise, and give advice on this book. Each of you have played a part in helping this book become a reality.

Lastly, I have had so many other mentors, friends, and family that have poured into me, helping me become the man I am today. I must thank each of you for your help in preparing me to write this book.

Foreword

"Just the Facts…" 20 years prior to the date of my birth was the date of the first episode of the TV show, *Dragnet*, in which Sgt. Joe Friday investigated clues and solved crime issues in one brief episode. In a world when answers to questions can seem more complex than they need to be and books addressing topics of interests are longer than what seems worth it, I am thrilled to introduce you to Josh and this book!

Josh is a wonderful human being who has chosen to write a book about the most wonderful Book—the Bible, and what it says about essential oils. In a desire to see Young Living Essential Oils used in every home, by an audience who places high value on the truth of scripture, Josh has succeeded in creating a valuable tool for existing and future essential oil users. As expected in a thorough book about essential oils, Josh faithfully explains the what's and how to's, but this book is special in that it also accurately examines several oils and evaluates them in their Biblical context. I believe the reader will find this exercise valuable, informative and important.

I told Josh this is the book I would write if I had the time or passion to write books. I believe it will serve as a valuable tool in the toolbox of many men and women who desire to enjoy the best of spiritual and physical health, wellness and abundance. It should be a welcome addition to every household as Young Living Essential Oils reach every home.

—*Adaryll Jordan, pastor and musician*

As an herbalist, I can say that learning about plants is a never-ending journey that delights the senses. The fact that God would pack so much in such tiny leaves, delicate and beautiful flowers, trees, barks, shrubs, roots, and stems shows how much God cares about each part of His creation. It still blesses me today. While I have read the Bible numerous times, I would generally speed over the plant names the same way I would over person's name or city that I can't pronounce. It wasn't until I started my formal studies into herbology that I began to notice the many plants mentioned in the Bible. It was like I was reading the Bible with fresh eyes.

This work is a unique take on plants, some of His oldest creations, as well as the how's and why's of using them effectively today. I love this book.

By utilizing the Hebrew and Greek translations, the expertise of rabbinical teachers, and even many of my own herbal books, Joshua was able to determine the particular species of plants the verses were actually referring to and using at that time in history (to the extent records allow). While it's not required that we use only the species referenced in the Bible, as many are no longer available today, having something that's as close as possible makes the scriptures come more alive for us today. How amazing is it to know that you're using an oil that is the same or very similar to the one Moses used!

As a Young Living member, I personally appreciate Young Living caring enough about staying as close to scripture as possible when they selected oils for their Oils of Ancient Scripture™ kit. Young Living uses amazing technology to identify and compare species on a DNA level to ensure the most similar match.

Whether you are new to herbs and essential oils, or you just enjoy a good Bible study, this book is for you. Both students of Biblical Greek and Hebrew and those with a vested interest in the topic will find this book thoroughly informative, inspiring, and enlightening. I know I have. If you enjoy teaching about oils from the Bible, this book is a great addition to your library and very different from what you may have already read.

Josh is indeed a consummate researcher of both the Bible and plants, as this book attests. I am fortunate to know him, having him as part of my YL family and mostly calling him my son-in-love, the husband of my daughter Sharon.

—*Debra Raybern, MH, ND, ICA*

Introduction

Many desire to completely follow the Lord, yet when it comes to wellness, people often discount the power of God's creation. II Timothy 3:16-17 states, "All Scripture is inspired by God and profitable for teaching, for reproof, for correction, for training in righteousness; so that the man of God may be adequate, equipped for every good work." This infers that everything in the Word works to build us into the men and women He desires us to be. Therefore, as we study His plants and their purpose, we become better equipped for every good work.

It was on the third day of creation that God established vegetation on the earth, including grass, fruit trees, herbs, and vegetables.

> *"Then God said, "Let the earth sprout vegetation, plants yielding seed, and fruit trees on the earth bearing fruit after their kind with seed in them" and it was so. The earth brought forth vegetation, plants yielding seed after their kind, and trees bearing fruit with seed in them, after their kind; and God saw that it was good. There was evening and there was morning, a third day."—Genesis 1:11-13*

Later on in this chapter of the Bible, God gives man dominion over all of the earth, thereby entrusting mankind to govern over the earth while exercising good stewardship. For us to fully accomplish God's command of stewardship over His creation, we must take time to carefully study these plants and their potential for enriching our lives.

Among the plants created and mentioned in the Bible, aromatic plants and their essential oils are referred to more than 200 times. In fact, they are strewn throughout the Bible from Genesis to Revelation. Both scientific research and experience corroborate, expand, and verify the possibilities of the many plants written about in ancient texts. Yet many people simply glance over them and never give much thought to these plants and their significance.

In this book, we are going to investigate several particular plants that are found in scripture (not just those found in The Oils of Ancient Scripture™ Kit). We are also going to learn about some essential oil blends whose ingredients include many Bible oils. We will begin our study by identifying the ancient use and significance of these various oils, and then move into their present and future value and purpose. You'll learn what essential oils are, how they can be used, and why they are important to our well-being.

Join me as we investigate all that God has gifted to us in His creation to support our journey to wellness.

Sketch of the royal seal of Hezekiah, king of Judah, discovered at the Temple Mount in Jerusalem dating back to 727-698 BC.

> The oils and aromatics mentioned in the Bible were considered more valuable than gold and silver. Israel's King Hezekiah kept "the spices, and the precious oil" (2 Kings 20:13) together with silver and gold in the royal treasure chamber. In ancient times, a person was considered truly wealthy if his/her treasure trove included essential oils.

What is an Essential Oil?

Essential oils are considered the "life blood" of the plant. 'Life blood' is not a scientific term, but just as blood in the human body flows throughout the body systems giving life, so do essential oils within a plant. More technically, they are liquid, natural compounds and constituents found in shrubs, flowers, and trees. Unlike fatty chain oils such as coconut oil or olive oil that absorb slowly into the skin, essential oils are light and volatile, able to be quickly absorbed into the skin due to their low molecular weight (size). Essential oils are very concentrated and one drop of oil represents the distillation of up to hundreds of pounds of plant material.

Ways to Use Essential Oils:

Topically
Oils can be applied neat (undiluted) to the skin or diluted with a carrier oil in a 1:1 or 1:2 ratio. Coconut, olive, and almond oil are great carrier oils to use when diluting oils for sensitive skin or children.

Aromatically
Oils can be inhaled directly from the bottle or diffused in a cold air diffuser made for therapeutic grade essential oils.

Internally
Some oils and oil blends are labeled for oral consumption, such as the Young Living Vitality™ line. These "Vitality™" labeled oils can be taken in capsules, added to drinking water and other beverages, consumed in a spoonful of organic honey or maple syrup, or added to food.

Historical Facts

Historical facts confirm that essential oils are not just a fad, but that they have in fact been used for centuries by many different people groups.

According to the University of Maryland Medical Center, "Essential oils have been used for therapeutic purposes for nearly 6,000 years. The ancient Chinese, Indians, Egyptians, Greeks, and Romans used them in cosmetics, perfumes, and drugs. Essential oils were also commonly used for spiritual, therapeutic, hygienic, and ritualistic purposes."[1]

- Essential oils are known to be the world's oldest natural medicine, referenced in both the Bible and other ancient texts as far back as 4500 BC.

- There are more than 600 references to plants mentioned in the Bible. Over 200 of these are known aromatics and many of them are specifically described as being in pure essential oil form.

- They are mentioned in Egyptian hieroglyphics as well as in ancient Chinese manuscripts.

- "Renowned writers from the Middle East such as Persian polymath Ibn Sīnā (980-1037 AD) and Islamic botanist and physician Ibn al-Baitair (1197-1248 AD) promoted the use of herbs and essential oils."[2]

- Soldiers used essential oils during both world wars.
- In 1922, more than 320 liters of pure essential oils were found in perfect condition in King Tut's tomb.
- The use of essential oils was revolutionized in Europe in 1937 by a French chemist named René-Maurice Gattefossé. This revolution helped to set the stage for modern day essential oil use.
- A leather satchel containing 65 vials of essential oils for creating perfume was discovered in the wreckage of the Titanic.
- There are over 8,000 references to essential oils in the National Library of Medicine.

Oils depicted in Ancient Egyptian Hieroglyphics.

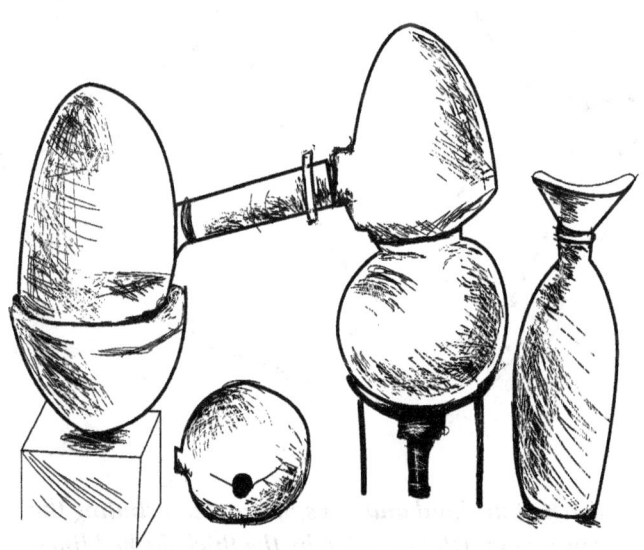

Terra cotta distilleries from around 3000 BC located in Pakistan Museum.

Ancient Spice Trade Route

The ancient spice trade flourished between the 10th century BC and the 5th century AD. During this time there were two major types of trade routes: the land route and the water route. Spices such as cinnamon, cassia, cardamom, ginger, turmeric, frankincense, myrrh, and sandalwood were loaded onto boats and camels to get to their destinations. These traders not only traded spices, but also ivory, silk, precious stones, gold, and other items. However, it was the spice trade that revolutionized the trading industry of ancient times, and many of these same spices are still traded to this day.[3]

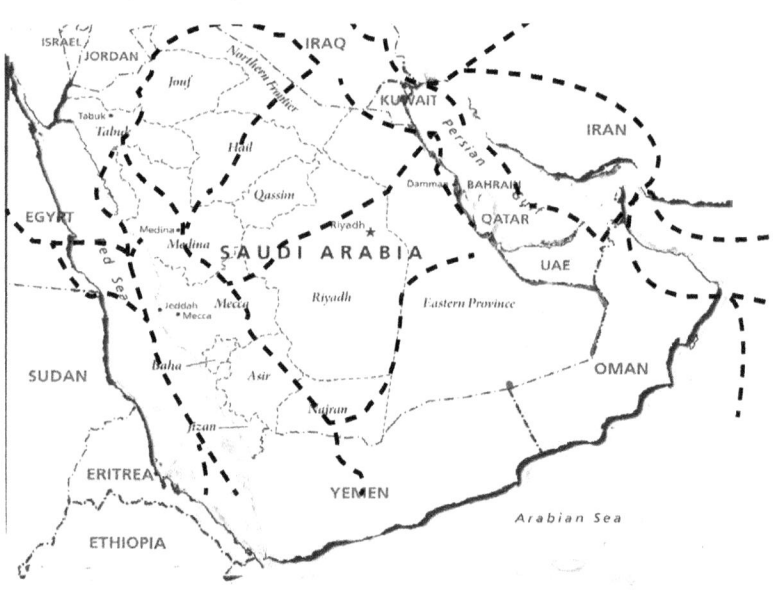

Sketch of the land and sea spice route, including the incense road. (Represented by the thick dashed lines.)

> For more information about this topic, numerous reference books and articles mentioned throughout this work are listed in the back of this book, both in the endnotes and the bibliography. (See page 99)

Usage in the Bible

Anointing

One of the most common ways essential oils were used in the Bible was for anointing. Per Vines Expository Dictionary, "anoint" as used in the Bible, means to "cover, rub or smear the head or body or object with oil, and in some cases it meant to pour over the head, body, or object."[4] Essential oils were used in the anointing of kings, for hospitality, on items reserved for sacred use, and to prepare bodies for burial. The essential oils used in the different types of anointing were usually mixed with olive oil; but flaxseed, walnut, sesame, and almond oil were also occasionally used.

Types of Anointing

Anointing Kings
- The act of anointing was very significant for dedicating people for the Lord's purpose. The anointing of a king was so powerful that it was equivalent to crowning him, as seen in the book of Samuel when Samuel anoints David as King (1 Samuel 16:13; 2 Samuel 2:4; II Kings 9:3-6).

Anointing For Hospitality
- Anointing was also an act of hospitality, which is one of the reasons why Mary anointed Jesus (Yeshua) (Luke 7:38, 46). It was Jewish custom for Jews to anoint themselves in order to refresh and invigorate their bodies. The Greeks and the

Romans anointed guests with oil and perfume at feasts. (Deuteronomy 28:40; Ruth 3:3; 2 Samuel 14:2; Psalms 104:15).

Anointing of Items For Sacred Use
- Moses anointed the tabernacle and all that was inside to sanctify it for use and visitation by the presence of God (Exodus 30:26-29, 40:9-11; Leviticus 8:10-11, and Numbers 7:1). This is just one example in the Old Testament of a physical thing or place being symbolically anointed and set apart for sacred use. In 2 Samuel 1:21 and Isaiah 21:5, the Israelite army anointed their shields in order to strengthen and consecrate them for Yahweh's protection in battle.

Anointing Body For Burial
- The bodies of the dead were sometimes anointed (Mark 14:8, 16:1; Luke 23:56).

Holy Anointing Oil
- The most sacred anointing oil was the Holy Anointing Oil, which was reserved for priests, kings, and sacred objects. Only holy men or priests were allowed to carry out the sacred task of anointing with this oil (Exodus 30:22-25).

Ancient Hebrew anointing a man.

Many feel that anointing a person or thing is reserved for priests or pastors, but anointing was actually a common practice among the Israelites and later among Christians in the early church. Through Christ (Yeshua), believers have been given direct access to God and have the opportunity to minister in His name. It is permissible for a believer to anoint his/her house and to pray God's protection over his/her home and family. Believers must use God's provisions to glorify His name and dedicate all things unto Him (Hebrews 7; Colossians 3:17; Matthew 6:17; Mark 16:1; James 5:14; Luke 7:38-50; Mark 6:13).

Perfume and Incense

Two additional common uses for essential oils in the Bible are for burning incense and making perfume. The offering of incense, or burning of aromatic substances, is a common component in the religious ceremonies of nearly all ancient nations (Egyptians, Babylonians, Assyrians, Phoenicians, Israelites, etc.). Similarly, the burning of incense held a prominent place in the tabernacle and temple-worship of Israel.

Perfume
- Traded from lands in the ancient Near East, spices were regarded as a luxury, to the point that the wealthy considered them a treasure. They were used in perfumes, fragrant incense, daily cooking, and in ointments for the body and grooming (Exodus 25:6, 30:23-25, 30:34-38; 1 Kings 10:10, 25; Ezekial 27:17-22).

Holy Incense
- Holy incense is not the same as the Holy Anointing Oil. As mentioned above, The Holy Anointing Oil was set aside for anointing priests, kings, and sacred objects. The holy incense was burned in the temple. "Three hundred and sixty-eight minas (405 pounds) of incense were prepared once a year in the temple; one mina for each day plus three minas to be used during the sacrifice on the Day of Atonement."[5] Some of the ingredients for this incense had to be specially prepared, and on the Day of Atonement, the high priest would enter the Holy of Holies and burn the holy incense before the altar as an offering unto God. The holy incense of the temple held the same restrictions as the Holy Anointing Oil; it was only to be used by a priest or holy man (Exodus 30:34-36).

Condiment, Ointment, and Wellness

Essential oils were also used as condiments in food, additions to ointments, and for body support and wellness for the Israelites and many of the surrounding nations of the Middle East. (In fact, many of the spices used in Israel were imported.)

Condiments For Food
- Just as in modern times, spices were used in everyday cooking. Mint was commonly used in the ancient world as a condiment, and spices such as cinnamon, cassia, cardamom, ginger, turmeric, and sandalwood were important items of commerce on the ancient spice trade routes (Matthew 23:23).

Ointments
- In ancient times, ointments were defined as oily substances. The typical recipe for an ointment would include an oil—generally olive or almond—various spices and herbs, and resins or waxes. The ointment was then sealed in a small alabaster box. Many of these ointments were very expensive and considered a luxury to be used exclusively by the wealthy. They were used for ceremonial and holy rituals, burials, as medicine, and in cosmetics. Many of these ointments have been noted as lasting hundreds of years, and some found in archaeological sites have retained their fragrance (Esther 2:12; Ruth 3:3; Matthew 26:12; Isaiah 1:6; Exodus 30).

Body Support and Wellness
- The Bible speaks frequently about health and wellness. The Levite priests, elders of the church, and physicians would use essential oils to support the health and wellness of the people (Numbers 19:6; Mark 6:13; James 5:14).

Bible Oils

Many Biblical oils are distilled and sold today for modern usage. Some of these are the exact same species of plant that would have been used in Bible times while others are similar species and would be considered equivalent to the original. In these cases, the original is often impossible to source, the plant is endangered, it no longer exists, or the plant type is still debated.

For example, a couple of the original Bible plant species have been put on the International Union for Conservation of Nature's red list of "most-threatened" plants; the Lebanon Cedar is one of these. Some countries have put restrictions on harvesting certain plants like Indian Sandalwood in India because they have been over-harvested in the past.

Although a handful of the Bible oils distilled today are equivalents, they are still an excellent representation of the amazing plant oils God has placed throughout scripture.

> **A note about plant species:** When God gave dominion and stewardship over His creation to man, it included protecting endangered plants, reforesting, and replanting when an entire plant is used. This ensures these plants are available for coming generations. This is the stance that Gary Young and Young Living take.

Exact Bible Oils

Bay

Bay oil is distilled mostly from the leaves of the tree.

Name: Bay *(Laurilus nobilis)*

Hebrew: רַעֲנָן (raanan) "green," אֶזְרָח (ezrach) "native" = green native tree growing in natural soil

Description: Laurus nobilis is an aromatic evergreen tree or large shrub with green, glossy leaves, native to the Mediterranean region. It is also known as bay laurel, sweet bay, bay tree, true laurel, Grecian laurel, laurel tree, or simply laurel. Laurus nobilis has a spicy, uplifting scent.

Verse: Psalm 37:35—"I have seen a wicked, violent man. Spreading himself like a luxuriant tree (bay tree) in its native soil."

Scriptural Background: Many scholars believe that Laurus noblilis is the bay tree, which is native to Israel. The particular structure of the words 'green' and 'native' only appear together once in scripture. The Apostle Paul also alluded to it in 1 Corinthians 9:25 and 2 Timothy 2:5 when he talked about the physical crown one could win in a race; history shows this crown would have been made out of the leaves of the bay tree.

Historical Info: Ancient Greeks used leaves of the laurel tree to crown their victors and scholars. The Chinese use it in wine and sweet foods. Many people today cultivate the laurel tree as an ornamental plant.

Modern Day Uses:
- Bay is commonly used to give an aromatic flavor to Mediterranean dishes.
- Young Living offers it as a Vitality™ Oil for internal use. (Vitality™ oils are oils that are approved for internal consumption.)
- Bay essential oil is soothing to the skin.
- It supports the digestive, nervous, respiratory, and lymphatic systems.

Cassia

Cassia is a cinnamon plant with the oil distilled mostly from the bark of the tree.

Name: Cassia *(Cinnamomum cassia)*

Hebrew: קְצִיעוֹת (qetsiah) "cassia" = a powdered bark, like cinnamon

Description: Cassia is an evergreen tree originating in southern China. It is widely cultivated there and elsewhere in southern and eastern Asia. Cassia is the most common and widely used type of cinnamon. Cassia is part of the same family as the cinnamon plant, and like cinnamon, the oil is distilled primarily from its bark.

Verse: Psalms 45:8—"All Your garments are *fragrant with* myrrh and aloes and cassia; out of ivory palaces stringed instruments have made You glad."

Scriptural Background: Cassia is mentioned three times in the Bible. The writer of Hebrews quotes Psalm 45:8 as evidence of Jesus (Yeshua) being God. This passage notes that His garments are anointed with the fragrant perfume spices of cassia, myrrh, and aloes (Hebrews 1:8-9). It is listed as one of the ingredients in the Holy Anointing Oil (Exodus 30:24), as well as one of the spices traded to the King of Tyre. Tyre was a Phoenician city that was known for its ports of trade. It is a part of modern day Lebanon (Ezekiel 27:19).

Historical Information: Cassia was traded during the spice trade that went through the Middle East during Biblical times. Cassia is one of the most powerful essential oils listed in ancient texts dating back some 1,700 years. It was very important to the Greeks and Romans, and is one of the major spices used in Chinese medicine. Although "its aroma is similar to cinnamon, cassia is chemically and physically quite different."[6]

Modern Day Uses:
- Dilution is recommended as cassia can be very "hot" when applied to the skin. Care must also be taken when diffusing cassia because it can also irritate the eyes. A great way to experience cassia is by diluting it 50/50 (one drop of cassia and one drop of carrier oil) and applying to the bottom of the feet.
- Cassia supports the immune, digestive, respiratory, and circulatory systems.

Cinnamon

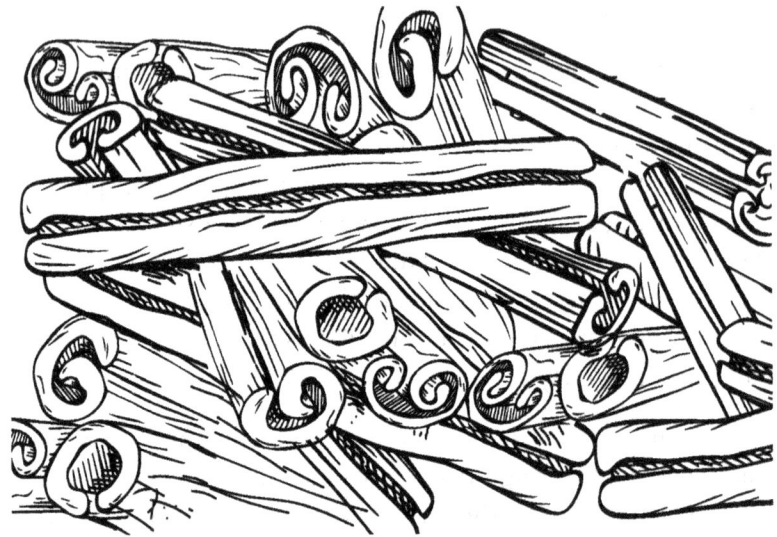

Cinnamon oil is distilled from the bark of the tree.

Name: Cinnamon *(Cinnamomum zeylanicum also known as Cinnamomum verum)*

Hebrew: קִנָּמוֹן (qinnamon) "cinnamon" = cinnamon bark

Greek: κινάμωμον (kinamómon) "cinnamon" = cinnamon

Description: *Cinnamomum verum*, also called true cinnamon tree (in order to distinguish it from cassia) or Ceylon cinnamon tree, is a small evergreen tree belonging to the family Lauraceae. This cinnamon is native to Sri Lanka and its leaves are oblong in shape. It has a characteristic odor of cinnamon and a very spicy, aromatic taste.

Verse: Exodus 30:23—"Take also for yourself the finest of spices: of flowing myrrh five hundred *shekels*, and of fragrant cinnamon half as much, two hundred and fifty, and of fragrant cane two hundred and fifty."

Scriptural Background: Cinnamon is mentioned four times in the Bible. Each of these occurrences addresses one of two issues: intimacy with God and/or intimacy between a man and woman.

Exodus 30:23 and Revelation 18:13 discuss man's relationship with God, and the oil is mentioned as one of the ingredients in the Holy Anointing Oil. Revelation 18:13 speaks about it being an item of commerce that need not be placed above worship of a holy God; here, God is calling His people out of sin.

Proverbs 7:17 and Song of Solomon 4:14 refer to the relationship between men and women. In Proverbs 7:17, cinnamon is used by the harlot to cover up the smells of indiscretion and to fill the room with a sweet smell in order to entice another young, unsuspecting man into unrighteousness. Song of Solomon 4:14 describes this spice as one that King Solomon's bride put on her robe for their wedding day.

Historical Info: Cinnamon was one of the major spices traded on the trade route through the Middle East. In fact, cinnamon was one of the spices that motivated world exploration. European explorers such as Christopher Columbus and Gonzalo Pizarro explored the world to find the source of cinnamon in order to keep up with the increasing demand for it. From this exploration, the New World was put on the map. Cinnamon was imported to Egypt as early as 2000 BC.

It was so valued in ancient times that it was considered a gift befitting royalty or deity. In fact, Emperor Nero of Rome burned cinnamon to honor his wife at her funeral. Today, it is used in common items such as toothpaste, gum, baked goods, and soda. It is one of the chief ingredients in the Young Living Thieves™ blend.

Modern Day Uses:
- Cinnamon can be diffused, directly inhaled, or diluted with V-6™ Oil Complex or another carrier oil and applied topically.
- It is commonly used to flavor dishes and is offered as a Vitality™ Oil.
- It supports the cardiovascular and digestive systems.

Coriander

Coriander oil is distilled from the seeds.

Name: Coriander *(Coriandrum sativum)*

Hebrew: גד (gad) "coriander" = coriander seed from its furrows

Description: Coriander is also known as cilantro or Chinese parsley. Coriander is an annual herb in the family Apiaceae, which is a family of mostly aromatic flowering plants. All parts of the plant are edible, but the fresh leaves and the dried seeds are the parts most traditionally used in cooking. Coriander has a sweet, warm fragrance that is calming and gently uplifting.

Verse: Numbers 11:7—"Now the manna was like coriander seed, and its appearance like that of bdellium."

Scriptural Background: Coriander is only mentioned two times in the Bible (Exodus 16:31; Numbers 11:7). In each occurrence, the Israelites compare manna to coriander.

Historical Info: Prized by the Egyptians, coriander was often placed in tombs, including the tomb of Ramses II. Coriander also had culinary usage during biblical times, as evidenced by traces of coriander found in pots at the Pre-Pottery Neolithic Archaeological Site in the Near East. Most coriander today is cultivated in eastern countries where the whole plant is used. It is also cultivated in the south of Europe, but here it is mostly cultivated for its seeds. These seeds are roughly the size and shape of a peppercorn.

Modern Day Uses:
- Dilute coriander with a carrier oil when using topically, then apply directly to target areas.
- Coriander is commonly used to flavor dishes, and is offered a by Young Living as a Vitality˝ Oil for this purpose.
- Coriander supports the digestive system and helps to normalize blood sugar levels.

Cypress

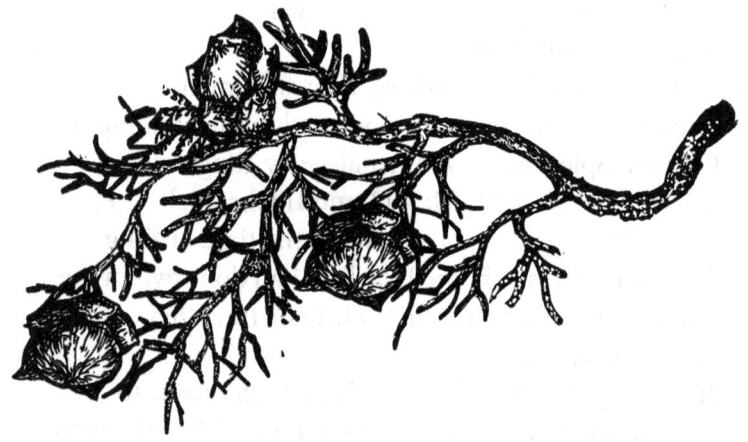

Cypress oil is distilled from the branches of the trees.

Name: Cypress *(Cupressus sempervirens)*

Hebrew: תִּרְזָה (tirzah) and בְּרוֹשׁ (berosh) and בְּרוֹתִים (beroth) and תְאַשּׁוּר (teashshur) "cypress" = a species of tree (Cypress)

Description: Cypress is a medium-sized coniferous evergreen tree that has a cone-like crown and loosely hanging, level branches.

Verse: Isaiah 44:14—"Surely he cuts cedars for himself, and takes a cypress or an oak and raises it for himself among the trees of the forest. He plants a fir, and the rain makes it grow."

Scriptural Background: Cypress is mentioned 22 times in the Bible and only in the Old Testament.
- After becoming king of both Judah and Israel, King David decides to move the Ark of the Covenant, and while this

was happening, David and the people celebrated with instruments made from cypress (2 Samuel 6:5).
- David writes in Psalm 104:17 that God takes care of His creation, and cypress is specifically mentioned as one of those regarded items.
- In I Kings and II Chronicles, cypress is mentioned as one of the main types of wood used in the building of God's temple and the king's palace by King Solomon (I Kings 5:8,10; 6:15, 34; 9:11 and II Chronicles 2:8, 3:5; Song of Solomon 1:17). (Between the time of King Solomon and Isaiah, the Israelites had done evil in the Lord's sight, and at the end of Isaiah's time they enter into Assyrian captivity.)
- In II Kings 19:23 and Isaiah 37:34, God speaks to Hezekiah regarding Sennacharib, king of Assyria, who was devastating the land and its people by over-harvesting the cedars and cypress trees.
- Nahum 2:3 speaks of spears made from cypress trees. These spears are used by the house of Israel to overthrow the ancient Assyrian city of Ninevah. In the midst of this, the people of Israel were beginning to cut down cedar and cypress trees to make idols for themselves, just as the Assyrians had, and they are condemned for doing this in Isaiah 44:14.
- God promises in Isaiah 41:19 and 55:13 that if Israel will turn back to Him as Yahweh and trust in His mercy, He will not only heal His people but will heal their land and the cypress, myrtle, cedar, acacia, olive, juniper, and box trees will fill the mountains again, which will be a visual memorial to the Lord's mercy.
- Hosea 14:8 refers to God as being a cypress that protects His people who come back to Him.
- In Ezekiel 27:5, Ezekiel laments the overthrow of the port city of Tyre by the Egyptians and mentions how their strong, beautiful ships were made from cypress.

The people of Tyre were known for their trade, and King Solomon obtained cypress for the temple from them.
- In Ezekiel 31:8, God warns Egypt of Assyria's fate by saying it stretches out further than a cypress, yet it succumbed to God's wrath. After this, Chaldean (Assyrian / Babylonian) King Nebuchadnezzar pushes the Egyptians back to their land and destroys Jerusalem, including the Jewish temple.
- In Isaiah 14:8, Isaiah prophesies that the Babylonians will no longer decimate the land by cutting down trees of cypress (which was a common practice of the Babylonians and Assyrians) once God finished pouring out His wrath on them. (Zechariah 11:2)
- Isaiah 60:13 states that the city of Zion (City of David—Jerusalem) will be rebuilt with cypress trees in order to beautify it, just like the temple King Solomon built.

Historical Info: Inscriptions were written on a Babylonian tablet made out of cypress dated to 1800 BC. The cypress tree is renowned for its durability. For example, the sturdy cypress doors of St. Peter's Basilica in Rome show no signs of decay, even after 1,200 years! This tree grows abundantly on the slopes of Mount Hermon in Israel. Its wood is hard, fragrant, very durable, and has a stimulating aroma. Some scholars think cypress is the gopher wood used in the construction of the ark.

Modern Day Uses:
- High in monoterpenes, cypress is excellent to apply topically after exercise to soothe achy muscles. Add to V-6™ Vegetable Oil Complex and apply for a soothing aromatic massage experience.
- Cypress is commonly used to support the circulatory and immune systems, to boost brain power, and in leg massages.

Dill

Dill oil is distilled using the whole plant.

Name: Dill *(Anethum graveolens)*

Hebrew: קֶצַח (qetsach) "dill" = fitches—seeds used as a condiment

Greek: ἄνηθον (anéthon) "dill" = dill

Description: Dill is an annual herb and is the sole species of the genus Anethum. It is in the celery family Apiaceae. Dill is native to Asia Minor and the Mediterranean. After many centuries of planting dill around the world, it now grows naturally in parts of Europe and North America.

Verse: Matthew 23:23—"Woe to you, scribes and Pharisees, you hypocrites! You pay tithes of mint, dill, and cumin, and have neglected the weightier provisions of the Law (Torah): justice and mercy and faithfulness; but these are the things you should have done without neglecting the others."

Scriptural Background: Dill is mentioned three times in scripture, twice in the Old Testament (Isaiah 28:25, 27) and once in the New Testament (Matthew 23:23).

In Isaiah, dill is one of the cultivated plants God mentions when warning Judah of what will happen if His people don't turn back to Him.

In Matthew 23:23, the Pharisees tithed it, and Jesus (Yeshua) rebuked them, based on Deuteronomy 14:22: "You shall surely tithe all the produce from what you sow, which comes out of the field every year." This illustrates that the paying of tithes with dill was in accordance with the Mosaic law (Torah), but the Pharisees were more concerned with paying tithes, and being noticed for paying tithes, than they were with actually sanctifying their lives to God.

Historical Info: Dill originated in southern Russia, the Mediterranean, and western Africa. It has been used as an herb for at least 5,000 years. Ancient Egyptians used it medicinally as an aphrodisiac and to repel witches. The Greeks used dill as a symbol of wealth, and the Romans believed that dill brought good fortune. The Romans also used dill leaves in the wreaths they made to recognize athletes and heroes. In ancient times, soldiers would apply burnt dill seeds to wounds to speed healing. In the Bible, dill is also mentioned as being used as condiment in cooking.

Modern Day Uses:
- Dill is believed to be calming when used aromatically. This oil can be blended with Roman chamomile to enhance a relaxing aromatic experience.

- Diffuse, directly inhale, or dilute with V-6™ Oil Complex to apply topically. Use caution to avoid possible skin sensitivity.
- Dill is commonly used to flavor dishes, and Young Living offers it as a Vitality™ Oil.
- It supports healthy blood sugar levels, as well as the digestive and nervous systems.

Frankincense

Frankincense oil is distilled from the resin found within the trunk of the tree.

Name: Frankincense *(Boswellia sacra and Boswellia carteri)*

Hebrew: לְבוֹנָה (lebonah) "frankincense" = frankincense

Greek: λίβανος (libanos) "frankincense" = frankincense or incense tree

Description: Frankincense is a small deciduous tree in the Burseraceae family. It has an odd number of leaves growing opposite each other along its branches. It has an earthy, balsamic scent.

Verse: Matthew 2:11—"After coming into the house, they saw the Child with Mary His mother; and they fell to the ground and worshiped Him. Then, opening their treasures, they presented to Him gifts of gold, frankincense, and myrrh."

Scriptural Background: Frankincense is mentioned 23 times in scripture.
- Frankincense was one of the ingredients used to make the Holy Incense (Exodus 30:34).
- It is a part of the grain offering submitted to God by the Israelite priests (Leviticus 2:1-2, 15-16; 6:15; 24:7 and Jeremiah 17:26; 41:5 and I Chronicles 9:29).
- It was not to be used with the sin offering (Leviticus 5:11 and Numbers 5:15).
- King Solomon and his bride use frankincense to describe their love for one another, as well as stating that it makes a good fragrance (Song of Solomon 3:6; 4:6,14).
- Since the institution of temple worship and King Solomon, much time had passed and the people of God had consistently done evil in God's, sight, including neglecting to make burnt offerings of frankincense (Isaiah 43:23). He stated that not even the offerings with frankincense were acceptable because they had been tainted (Isaiah 66:3). Jerusalem's destruction is prophesied, and the Israelites go into captivity (Jeremiah 6:20).
- During the Persian captivity, the people of God were allowed to rebuild the temple. One of the priests allowed a foreigner to sleep in the room where the grain offering was prepared and the frankincense was stored. This was not allowed, so the temple had to be cleansed and all items for the grain offering returned (Nehemiah 13:5, 9).
- Isaiah prophesies that Zion will be restored and that the people who had been scattered would bring frankincense with them (Isaiah 66:3).
- Frankincense was brought to Jesus (Yeshua) by the magi when He was 2 years old (Matthew 2:11).

- Revelation 18:13 mentions frankincense as an item of commerce that need not be placed above worship of a Holy God.

Historical Info: No other oil has been studied more or used by more cultures than frankincense. It has been traded for over 5,000 years and was very important to the Babylonians, Assyrians, Egyptians, and Israelites.

There are two major species of frankincense, both of which are offered as essential oils: *Boswellia carteri*, of African origin, and *Boswellia sacra*, of Arabian origin. The Bible does not tell us which species of frankincense was given to Jesus (Yeshua,) leaving this to be commonly debated. This background research sheds light on the question: "The four ancient Israeli towns of Haluza, Mamshit, Avdat and Shivta, along with associated fortresses and agricultural landscapes in the Negev Desert, are spread along routes linking them to the Mediterranean end of the incense and spice route. Together, they reflect the hugely profitable trade in frankincense and myrrh from southern Arabia to the Mediterranean, which flourished from the 3rd century BC until the 2nd century AD."[7]

Israel was at the northern end of the trade route and the Israelites desired to control it. The magi came from a region to the east that was a part of the Arabian spice trade land route. *Boswellia sacra* was traded through this route, so it is most probable that this is the species that would have been brought to Jesus (Yeshua). Additionally, it was the species most sought after in ancient times. Chemically, *Boswellia carteri* and *Boswellia sacra* are the most closely related of the four species of frankincense. Collectively, they are the two most potent species of frankincense, and they support the same body systems.

Modern Day Uses:
- Diffuse or simply smell during prayer time.
- In addition to these uses, frankincense contains the naturally occurring constituent alpha-pinene, which is known to help in the appearance of healthy-looking skin.
- It also has an ingestible Vitality™ Oil.
- It supports the respiratory, lymphatic, and nervous systems, and it also helps with healthy brain function.

Galbanum

Galbanum oil is distilled from the resin derived from the stems and branches.

Name: Galbanum *(Ferula gummosa)*

Hebrew: חֶלְבְּנָה (chelbenah) "galbanum" = galbanum, an odorous gum

Description: Galbanum has a spicy, woody, balsam-like fragrance; some believe it to be pungent. It is an evergreen bush that is native to Asia and grows abundantly in the mountains of Iran. It has white flowers that grow into orange fruits.

Verse: Exodus 30:34—"Then the Lord said to Moses, "Take for yourself spices, stacte and onycha and galbanum, spices with pure frankincense; there shall be an equal part of each."

Scriptural Background: Galbanum is only mentioned one time in the Bible as a part of the incense blend burned by the high priest in the Holy of Holies on the Day of Atonement.

Historical Info: Galbanum was used by the Greek physician Hippocrates and the Roman herbalist Pliny the Elder. It was very precious and sacred to the Egyptians, who used it as resin and incense. In fact, it was one of the major components of the Egyptian perfume trade. It has a long history of use in skin treatments and for prolonging the life and scent of other oils. The Mesopotamians loved it so much, they called it 'Mother Resin.' It is currently used in cosmetics, food flavoring, art, manufacturing varnish, and to support various body systems.

Modern Day Uses:
- Galbanum may be inhaled or applied topically.
- It may reduce the appearance of fine lines and wrinkles on the skin.
- It is great for use in a leg massage.
- It supports the nervous, digestive, and exocrine systems.

Hyssop

Hyssop oil is distilled from the stems and leaves of the plant.

Name: Hyssop *(Hyssopus officinalis)*

Hebrew: אֵזוֹב (ezob) "hyssop" = hyssop

Greek: ὕσσωπος (hussópos) "hyssop" = hyssop

Description: Hyssop has lavender-like blooms that are soft when cut fresh and still full of their wonderful slightly sweet, minty aroma. While the blooms are minty, the leaves are quite bitter. It is native to southern Europe, the Middle East, and the region surrounding the Caspian Sea.

Verse: Psalm 51:7—"Purify me with hyssop, and I shall be clean; wash me, and I shall be whiter than snow."

Scriptural Background: Hyssop is mentioned 12 times in the Bible.

Hebrews 9:19-20 talks about the old covenant made between God and man in Exodus 24. To affirm the covenant, Moses took blood, water, scarlet wool, and hyssop, and sprinkled it all over the people, the tabernacle, and the book of the law (Torah). This blood sacrifice cleansed the people and gave atonement for their sins.

Hyssop was used in cleansing rituals in Leviticus 14:4, 6, 49, 51, 52 and Psalm 51:7.

In Numbers 19:6 and 19:18, God commanded the Israelites to dip hyssop stalks into the passover lamb's blood and strike the doorposts of their homes to protect them from the plague of the death angel. The oils in the hyssop branches were released into the air and also mixed with the blood. The death angel then passed over their home as every first born of the Egyptians was killed.

It was also one of the plants about which King Solomon received wisdom (I Kings 4:33).

Hyssop was offered to Jesus on the cross (John 19:29). Jesus' (Yeshua's) death on the cross fulfilled the blood sacrifice (old covenant) in Exodus, thereby ringing in a new covenant, wherein Christ's sacrifice became sufficient for the atonement of sins. Jesus (Yeshua) was and is the promised Messiah. Jesus' (Yeshua's) final drink, which contained hyssop, bore the symbolism of His death and the cleansing of sin.

Historical Info: Hyssop was considered a sacred oil in ancient Egypt, Israel, and Greece. The Hebrews called hyssop a holy herb because it was used to cleanse sacred places. Additionally, the ancient Romans and Greeks valued hyssop and used it to support the respiratory system. Three hundred years ago, hyssop tea and tinctures were used to support different body systems. The plant is also commonly used by beekeepers to produce a rich and aromatic honey. Due to its intensely bitter taste, it is used only moderately in cooking. More notably, it is used to flavor liqueur, such as the official formulation of Chartreuse that combines distilled alcohol and 130 herbs, plants, and flowers, including hyssop.

Modern Day Uses:
- Hyssop aids the body's natural response to irritation and injury.
- It has natural cleansing and purifying properties.
- Hyssop supports the respiratory and digestive systems.

Mint "Peppermint"

Mint "Peppermint" oil is distilled from the leaves, stems, and flower buds.

Name: Mint "Peppermint" *(Mentha piperita)*

Greek: ἡδύοσμον (héduosmon) "mint or peppermint" = sweet smelling plant (mint)

Description: Peppermint leaves are broad and dark green with reddish veins. The leaves and stems are usually slightly fuzzy. Peppermint is a fast-growing plant; once it sprouts, it spreads very quickly and grows best in shaded, moist areas. It is a naturally occurring hybrid of spearmint and water mint.

Verse: Matthew 23:23—"Woe to you, scribes and Pharisees, you hypocrites! You pay tithes of mint, dill, and cumin, and have neglected the weightier provisions of the Law (Torah): justice and mercy and faithfulness; but these are the things you should have done without neglecting the others."

Scriptural Background: Mint is only mentioned twice in scripture, both times in the New Testament. Mint was a widely-cultivated plant in Bible times. In Matthew 23:23 and Luke 11:42, the Pharisees tithed it and Jesus (Yeshua) was angered, for Deuteronomy 14:22 states, "You shall surely tithe all the produce from what you sow, which comes out of the field every year." While tithing with mint was in accordance with the Mosaic law (Torah), the Pharisees were more concerned with being noticed for paying tithes than they were with actually sanctifying their lives to God.

Historical Info: Three common kinds of mint used in the Bible times were horse mint, garden mint, and peppermint. As the Greek translation for the word mint refers to "peppermint" specifically, it is likely that the most common form was peppermint. It is thought to have originated in northern Africa and the Mediterranean. It is listed in the Ebers Papyrus, an ancient Egyptian medical text dating to 1550 BC. Mint was so valued in Egypt that it was used as a form of currency.

Peppermint provides a great example of the potency of essential oils: one drop of peppermint oil is roughly equivalent to 26 cups of dried herb peppermint tea. (This is one reason many people who once used traditional herbal formulas are switching to essential oils. People are learning that a small amount of essential oil goes a long way, and oils are many times more powerful than their dried or tinctured herbal counterparts.)

Modern Day Uses:
- The scent of Peppermint invigorates the mind and senses, and it may enhance focus and concentration.
- Peppermint oil creates a cool, tingling sensation on the skin, making it a favorite for sports massage. It is considered a "hot" oil, so it is sometimes necessary to dilute for topical application on individuals with sensitive skin.
- It is commonly used as an ingredient in food. Peppermint Vitality™ is specifically labeled for internal usage.
- It supports the digestive and respiratory systems.

Myrrh

Myrrh oil is distilled from the resin of the tree.

Name: Myrrh *(Commiphora myrrha)*

Hebrew: מֹר (mor) and לֹט (lot) "myrrh" = myrrh

Greek: σμύρνα (smurna) and σμυρνίζω (smurnizo) "myrrh" = myrrh, a bitter gum

Description: Myrrh is made from the resin of dried tree sap. It is a flowering tree that is native to the Arabian Peninsula and Africa, most notably to Oman, Yemen, Djibouti, Ethiopia, Somalia, and northeast Kenya. It belongs to the botanical family Burseraceae, the same plant family as frankincense.

Verse: Esther 2:12—"Now when the turn of each young lady came to go in to King Ahasuerus, after the end of her twelve months under the regulations for the women—for the days of their beautification were completed as follows: six months with oil of myrrh and six months with spices and the cosmetics for women."

Scriptural Background: Myrrh is mentioned 16 times in the Bible.
- In Genesis 37:25 Joseph, son of Jacob and Rachel, was sold to Midianite traders by his brothers. These traders brought him to Egypt. They were carrying spices such as aromatic gums, balm, and myrrh.
- Later, in Genesis 43:11, Joseph was made second in command over Pharaoh's house and there was a famine in the land. When Joseph's long-lost family was forced to travel to Egypt to search for food, they met with Joseph but did not recognize him. He offered conditional help if the brothers agreed to return with the youngest brother, Benjamin. When they returned to their father and told him of Joseph's request, Jacob was fearful for Benjamin, but sent him to Joseph with a present of fruit, balm, honey, spices, myrrh, nuts, and almonds.
- Myrrh is mentioned as one of the main ingredients in the Holy Anointing Oil (Exodus 30:23).
- In Esther 2:12, Esther underwent six months of purification and beautification with myrrh oil.
- In Psalms 45:8, it is used by the writer of the book of Hebrews as evidence of Jesus' (Yeshua's) divinity, and this book also mentions that His garments are anointed with the fragrant perfume spices of cassia, myrrh, and aloes (Hebrews 1:8-9).
- In Proverbs 7:17, it is used by the harlot to cover up the smells of indiscretion and to fill the room with a sweet smell in order to entice another young, unsuspecting man.

- King Solomon uses myrrh in reference to fragrance and perfumes used by him and his beloved to show their intimate love for each other (Song of Solomon 1:13; 3:6; 4:6; 4:14; 5:1; 5:5; 5:13).
- It is one of the gifts the magi presented to Jesus (Yeshua) when he was around 2 years old (Matthew 2:11).
- Myrrh and sour wine were offered to Jesus (Yeshua) before He was put on the cross (Mark 15:23).
- When Jesus (Yeshua) reached Golgotha, He was first offered sour wine and gall (which He spit out—Matthew 27:34), then sour wine and myrrh (but He would not take it—Mark 15:23). While he was on the cross, He was offered sour wine and hyssop (Matthew 27:48 and John 19:29).
- Nicodemus took care of Jesus' (Yeshua's) body after He was crucified and bound His body with linens and applied spices of myrrh and aloes (John 19:39).

Historical Info: Myrrh and frankincense were key trade components throughout the Middle East and northern Africa for more than 5,000 years, and were used by the Babylonians and the Assyrians for religious ceremonies. The Egyptians used both oils for skin treatments, insect repellent, perfume, and incense.[8] Myrrh was used to rejuvenate the skin, and frankincense was used in powder form as makeup. Both were found pictorially represented on the walls of Queen Hatshepsut's temple. Ruler of Egypt from approximately 500-480 BC, Queen Hatshepsut was considered one of the most beautiful women in the world, and history shows that myrrh clearly played a role. Additionally, myrrh was often used to prepare a woman for labor as well as for use after her baby's birth.

Modern Day Uses:
- Myrrh prolongs the life and scent of other oils.
- It may reduce the appearance of fine lines and wrinkles on the skin.
- It supports the exocrine and immune systems.

> Hint: since myrrh is one of the more viscous oils, applying a small amount of olive oil to the opening of the bottle, the edge of the reducer, and the inside of the bottle cap is suggested. This will prevent the cap from sticking and becoming difficult to open over time.

Myrtle

Myrtle oil is distilled from the leaves of the tree.

Name: Myrtle *(Myrtus communis)*

Hebrew: הֲדַס (hadas) "myrtle" = myrtle tree

Description: The common myrtle is an evergreen shrub native to southern Europe, northern Africa, western Asia, Macaronesia, and the Indian subcontinent. It has a fresh, eucalyptus-like scent.

Verse: Nehemiah 8:15—"So they proclaimed and circulated a proclamation in all their cities and in Jerusalem, saying, "Go out to the hills, and bring olive branches and wild olive branches, myrtle branches, palm branches and branches of other leafy trees, to make booths, as it is written."

Scriptural background: Queen Esther's Hebrew name "Hadassah" means "myrtle." Myrtle is mentioned six times in scripture.

God promised in Isaiah 41:19 and 55:13 that if captive Israel would turn back to Him as Yahweh and trust in His mercy, He would not only heal His people but would heal their land. He also promised the cypress, myrtle, cedar, acacia, olive, juniper, and box trees would fill the mountains again, which would be a visual memorial to the Lord's mercy.

In Nehemiah 8:15, Ezra reads from the book of the Law (Torah) and reinstates the Feast of Booths which is mentioned in Leviticus 23:40. It states: "Now on the first day you shall take for yourselves the foliage of beautiful trees, palm branches and boughs of leafy trees and willows of the brook, and you shall rejoice before the Lord your God for seven days." Myrtle was one of the main plants used in the reinstatement of the Feast of Booths (Nehemiah 8:15).

In Zechariah 1:8,10, and 11, the prophet Zechariah has a prophetic vision where an angel of the Lord is standing in the midst of myrtle trees and proclaims God's promise to restore the Israelites back to Jerusalem one day. Myrtle was one of these promised trees.

Historical Info: Myrtle was a prominent plant in Roman gardens, especially in those of the elite. Myrtle was attributed to the mythological goddess Venus. Additionally, Queen Victoria of England had myrtle in her wedding bouquet, and a sprig of it was planted at the Osborne House in the United Kingdom. In modern times, myrtle from that very sprig is used in the royal weddings.

Modern Day Uses:
- With a soothing effect when inhaled, myrtle is said to promote overall well-being.
- Myrtle encourages proper nasal function and is often used in place of peppermint or eucalyptus with young children.
- It is mild when applied to the skin, so dilution is not typically necessary.
- Myrtle has been studied for its supportive and soothing effects on the respiratory system.
- Myrtle encourages healthy skin and hair.

Spikenard

Spikenard oil is distilled from the roots of the plant.

Name: Spikenard *(Nardostachys jatamansi)*

Hebrew: נֵרְדְּ (nerd) "nard"—nard or spikenard

Greek: νάρδος (nardos) "spikenard"—spikenard or nard

Description: Known as true nard, it is a flowering plant with pink, bell-shaped flowers, found in the Himalayan mountains of Nepal, China, and India.

Verse: John 12:3—"Mary then took a pound of very costly perfume of pure nard, and anointed the feet of Jesus (Yeshua) and wiped His feet with her hair; and the house was filled with the fragrance of the perfume."

Scriptural Background: Spikenard, or nard, is mentioned five times in the Bible. Just like cinnamon, each reference involves one of two issues: intimacy with God and/or intimacy between a man and woman.

Song of Solomon 1:12 and 4:13,14 speaks of an intimate relationship between a man and woman. These verses display this spice as one that King Solomon used to express his deep love for his bride.

Mark 14:3 and John 12:3 address mankind's relationship with God and His son, Jesus (Yeshua). In this passage, Mary anoints Jesus (Yeshua) for burial by using a very precious oil, one that would not typically be used to anoint guests. This oil was so precious during those days that Judas Iscariot, the disciple who managed the money, replied "why was this perfume not sold for three hundred denarii and given to poor people?" That amount was roughly equivalent to a couple years' average salary of that time. In fact, before Mary anointed Jesus (Yeshua) with the perfume, she had it safely sealed in an alabaster box to prevent evaporation. This shows how precious and highly sought after this oil was in ancient times.

Historical Info: Spikenard was, and still is, a very expensive oil. To the Romans, it was a highly prized perfume as well as a frequent ingredient in cooking. Nard is mentioned in an encyclopedia written by Pliny the Elder called Natural History, which was written sometime between AD 77 and AD 79. It was found in perfume form in the tomb of the Egyptian pharaoh Tutankhamen, and was an ingredient of the incense used in Egyptian temples. It was also used as a seasoning by the Europeans in the Medieval Times and as a medicinal herb in India. Pope Francis' coat of arms has a nard flower on it.

Modern Day Uses:
- Spikenard can be used aromatically or topically.
- It is supportive of healthy skin.
- It also supports the nervous and circulatory systems.

Indian Sandalwood

Sandalwood oil is distilled from the wood of the tree.

Name: Indian Sandalwood *(Santalum album)* also known as Sacred Sandalwood

Hebrew: אַלְגּוּמִים (algummim) "algum" and אַלְמֻגִים (almuggim) "almug", עֵץ (ets) "wood" = wood from the algum or almug tree

Description: Sandalwood is an evergreen tree native to Asia. The tree grows to a height of about 30 ft. and has small purple flowers and small fruits containing seeds. There are many species of sandalwood.

Verse: I Kings 10:11—"Also the ships of Hiram, which brought gold from Ophir, brought in from Ophir a very great number of almug trees and precious stones."

Scriptural Background: The algum tree is mentioned six times in the Bible. The Hebrew word "Algummim" is translated to Algum or Almug tree, and can be found in I Kings 10:11, 12a, 12b and II Chronicles 2:8, 9:10, 11. Wood from this tree was used to build the pillars of the temple King Solomon was building and for the musical instruments used there.

It is thought that the city of Ophir mentioned in these Bible passages is in southern India, which was a part of the shipping trade route of the ancient world. This area cultivated paddy, sugarcane, millets, pepper, various pulses, coconuts, beans, cotton, plantain, tamarind and sandalwood. It is even thought that there may have been a land trade route from southern to northern India. It is known that sandalwood, precious stones, spices, and several types of produce were traded along this route. Two types of sandalwood are found in this area: red sandalwood and Indian sandalwood. Both types were frequently used in woodworking. It is most likely that Indian sandalwood is the Algum tree of the Bible.

Historical Info:
In ancient times, sandalwood was a very valuable tree to both the Chinese and the Phoenicians, and today it remains a very important and expensive wood to acquire. Its rarity has not discouraged its use or significance for the nation of India. In ancient times, they used it to make furniture, caskets, incense, and canes, and many modern Indian dishes call for sandalwood as a spice. This flavorful spice is also widely used in Europe.

Many of these tree species are closely monitored by the Indian government in order to conserve them and protect them from over-harvesting. As a result, they can be difficult to acquire.

Young Living has partnered with a farm in Australia that has successfully transplanted the Indian Sandalwood.

Modern Day Uses:
- Sandalwood promotes a deep sense of relaxation and emotional well-being.
- It enhances deep sleep by stimulating the release of natural melatonin.
- Sandalwood is often used for brain power, clarity of thought, and greater focus. Smelling sandalwood promotes relaxation and inspires spiritual awakening.
- Sandalwood helps maintain the appearance of healthy, beautiful, youthful, toned skin.
- It is supportive of overall health and well-being.

The Indian Sandalwood *(Santalum album)* species is closely monitored by the Indian government in order to conserve and protect them from over-harvesting. Since the Indian sandalwood is so scarce, Royal Hawaiian Sandalwood *(Santalum paniculatum)* is often used. It was a major item of trade in Hawaii in the early 1800s because of the Panic of 1819 when it was hard to obtain sandalwood and other goods from China and the surrounding areas. The wood was used as currency, incense, and for use in woodworking. In the mid-1800s, the sandalwood trade in Hawaii ceased because of over-harvesting and debt. In recent years, the Hawaiian sandalwood has begun to make a come back as a viable equivalent for the difficult-to-find Indian Sandalwood.

Equivalent Bible Oils

Ceder

Cedarwood oil is distilled from the bark of the tree.

Name: Ceder "Cedar of Lebanon" *(Cedrus libani)*

Hebrew: אֶרֶז (erez) "cedar" and אַרְזָה (arzah) "cedarwork" and אָרַז (araz) "made of cedar" עֵץ "wood or tree" = wood from the cedar tree

YL Equivalent: Cedarwood *(Cedrus atlantica)*

Description: Cedarwood is also known as the Atlas Cedar, and is a large, coniferous evergreen tree. It is native to the Atlas Mountains of Morocco and Algeria. It is considered by many to be a subspecies of the Lebanon cedar.

Verse: Leviticus 14:49—"To cleanse the house then, he shall take two birds and cedar wood and a scarlet string and hyssop"

Scriptural Background: Cedar is mentioned 73 times in the Bible.
- Job describes the behemoth who has a tail like a cedar (Job 40:16).
- Cedar is used to cleanse houses in the Old Testament (Leviticus 14:4, 6, 49, 51, 52 and Numbers 19:6).
- In Numbers 24:6, Baalam prophesies a blessing over Israel, and his blessing includes cedars.
- Jotham uses cedars in a metaphor to the people (Judges 9:15).
- David and King Hiram of Tyre traded between each other, and David built his own home out of these cedars (II Samuel 5:11; 7:2 and I Chronicles 14:1; 17:1).
- David writes in Psalm 104:17 that God takes care of His creation, and both cypress and cedar are specifically mentioned as two of those regarded items (Ezekiel 17:22, 23 and Psalm 80:10).
- God saw that David built a house of cedar for himself, yet there was no home for Him. Therefore, He commanded David to build Him a house out of cedar (II Samuel 7:7 and I Chronicles 17:6).
- During his time as king, David sinned against the Lord, and God told him that he would not finish building the temple as a result, rather his son Solomon would. King David stocked up on the supplies, including cedars, that Solomon would need to complete the temple (I Chronicles 22:4a, 4b; 28:2-3, 6).
- King Solomon continued the positive trade relationship/ alliance with the king of Tyre that his father started and continued to receive cedar and cypress and many other goods from him to complete the temple (I Kings 5:6, 8,10; 9:11 and II Chronicles 2:3, 8).
- Solomon asked God for wisdom, and God gave it to him. Cedar is one of the trees about which God gave him wisdom, and Solomon transplanted cedar trees from the

king of Tyre with which to fill the land. (King Solomon was not only king; he was also a horticulturist.) (I Kings 4:33; 10:27 and II Chronicles 1:15; 9:27).
- King Solomon began the temple construction using cedar. He used cedar as the planks, then placed cypress, and overlaid them with gold. In fact, almost everything in the temple was built in this manner, for cedar is a wood that is bug resistant, does not decay, and is very durable (I Kings 6:9, 10, 15, 16, 18, 20, 36; 7:2a, 2b, 3, 7, 11, 12).
- In Song of Solomon 1:17, 5:15, and 8:9, Solomon and his bride use cedar in their metaphors of love towards each other (metaphors of strength and sturdiness).
- King Jehoash of Israel sent a message to King Amaziah of Judah stating that Israel would defeat Judah if they met. This message was in the form of a metaphor referencing cedar (II Kings 14:9 and II Chronicles 25:18).
- In response to the evil acts of children of Israel, the prophet Amos prophesied that God was going to rain down judgment upon both Judah and Israel. He stated that even though the Amorites were tall like the cedars and strong like the oak, His judgment had rained down hard on them, and that Israel too would feel such judgment (Amos 2:9).
- In Isaiah 14:8, Isaiah prophesied that the Babylonians would no longer decimate the land by cutting down trees of cypress and cedar, which was a common practice of the Babylonians and Assyrians, once God had finished pouring out His wrath on them. (Zechariah 11:1-2)
- In II Kings 19:23, II Chronicles 25:18, and Isaiah 37:34, God spoke to Hezekiah regarding Sennacharib, king of Assyria, who was devastating the land and its people by over-harvesting the cedars and cypress trees. God promised to defeat a portion of the Assyrians, which He later did. Then Zephaniah prophesied that God would lay

waste to the Assyrians, just as they had laid waste to the forest of cedars (Zephaniah 2:14).
- The prophet Jeremiah prophesied that the Israelites would again fall to another adversary who would destroy cedars (Jeremiah 22:7, 14, 15, 23).
- Cedar is again mentioned in the parable of two eagles and a vine in Ezekiel 17:3.
- In Ezekiel 27:5, Ezekiel lamented the overthrow of the port city of Tyre by the Egyptians, mentioning how their strong, beautiful ships were made out of cypress and cedars.
- The people of Tyre were known for their trade and are remembered as the people from whom King Solomon received cypress and cedar for the temple construction (Ezekiel 27:24).
- In Ezekiel 31:3, 8, God warned Egypt of Assyria's fate by saying it was once beautiful like cedars of Lebanon and stretched out further than a cypress, yet succumbed to God's wrath, making it known that God was more powerful than His creation (Psalm 29:5).
- God's people became arrogant with His promise to restore the land, and they claimed cedars as a thing to be restored by their hands instead of God's (Isaiah 9:10).
- God stated in Isaiah 2:13 that His wrath would be against all of creation including cedars; for many nations—including Israel at times—made idols to other gods with cedar, oak, and cypress (Isaiah 44:13-17).
- Creation is to praise the Lord and His creation; even cedars will praise Him, and a righteous man will grow like cedars of Lebanon (Psalm 148:9).
- The book of Ezra describes that both the people of Israel and Judah were permitted to come back to Jerusalem during the Persian captivity to rebuild the city and temple with cedars from the people of Tyre (Ezra 3:7).

Historical Info: In ancient times, cedar was used in cleansing rituals for lepers and after touching a "dead thing". As cedarwood was used by many cultures more than 5,000 years ago, it is possibly the first distilled oil. In fact, Lebanon Cedar was a prized wood for Israel, Egypt, and the Phoenicians. It was used in temple, palace, and in boat construction. The Egyptians and the Greeks burned it in their temples. Today, fences are often made of cedar, as well as cedar closets and hope chests that protect clothing from moths and other damaging insects.

Because of extreme harvesting, it is estimated that only 5% of the Lebanese cedar forest is left in Lebanon and Syria. It is on the International Union for Conservation of Nature (IUCN) red list of most threatened trees. For this reason, Young Living uses the Atlas Cedar as its equivalent. These two species are so remarkably similar that when the tissue of both the Atlas Cedar and the Lebanon Cedar were tested, no distinguishing gene markers were detected between the two species.

Modern Day Uses:
- Cedarwood may enhance a restful night when a drop is placed on the corners of a pillow, and it can promote mental clarity when inhaled.
- Cedarwood has been historically recognized for its calming and purifying properties.
- Cedarwood can also be used to assist with overcoming emotional blockages that may prevent one from moving forward.
- Cedarwood may be diffused, inhaled, or worn topically. Dilution is generally not needed. A couple drops of cedarwood and frankincense massaged into the scalp produce a truly divine aroma and shiny, manageable hair.
- It also supports the lymphatic and immune systems.

Juniper

Juniper oil comes from the bark of the tree.

Name: Juniper or Broom—plant type is highly debated see note on page 76*

Hebrew: רֶתֶם (rethem or rothem) "juniper or broom" = juniper tree or broom bush.

YL Equivalent: Juniper *(Juniperus osteosperma)*

Description: Juniper is a shrub or tree native to the southwestern United States and is also known as Utah Juniper. It grows in dry soil and has thick shoots and an earthy, woodsy aroma.

Verse: I Kings 19:4—"But he himself went a day's journey into the wilderness, and came and sat down under a juniper tree; and he requested for himself that he might die, and said, 'It is enough; now, O Lord, take my life, for I am not better than my fathers.'"

Scriptural Background: Juniper, or "broom tree," is only mentioned four times in scripture.

In I Kings 19:4-5 Elijah killed the prophets of Baal, and then Jezebel threatened to kill him. Elijah fled to the wilderness towards Beersheba. On his journey, he rested under a Juniper tree, and there an angel of the Lord provided food for his journey.

In Job 30:4, Job used juniper as food.

In Psalm 120:4, David described burning coals of the juniper tree as a metaphor of sin and its consequences, for the coals of broom/juniper were known to burn very hot and very long.

Historical Info: The Juniper tree has been used for centuries as food, for supporting body systems, for fuel, and for its wood (used for making shelter, utensils, and furniture). The ancient Egyptians, ancient Greeks, and Native Americans used it to support healthy body systems. Romans used it in cooking. During Medieval times, bundles of juniper berries were hung over door posts and sometimes buried next to entrances to ward off witches. It is the main flavor in the European alcoholic beverage, gin. In India, juniper leaves are burned as incense.

Modern Day Uses:
- Juniper oil can be diffused or used topically. Dilution with a carrier oil may be necessary for topical application, as juniper is considered a "hot" oil.
- It supports mental clarity and memory, and when diffused, it has a relaxing aroma.
- It supports the digestive, urinary, and nervous systems.

> It is still debated amongst botanists, scholars, and theologians as to which plant species is actually referred to in this passage. So much time has distanced us from the source that many things are subject to speculation. In many cases, the Hebrew cannot be accurately translated, even by linguistic experts. The Hebrew word most probably means "juniper tree" or "broom bush," a bush found in the deserts of Syria and Arabia, but it is unclear as to what that is. The three most probable choices are: the Phoenician Juniper *(Juniperus phoenicea)*, the Spanish Broom *(Spartium junceum)*, or the white broom *(Retama raetam)*. These are all three native to Israel, Arabia, and several other parts of the Middle East. They all grow best in the desert.

Onycha

Onycha oil is distilled from the resin of the tree.

Name: Onycha—meaning is highly debated see note on page 79*

Hebrew: שְׁחֵלֶת (shecheleth) "onycha"—snail or a flower

YL Equivalent: Onycha *(Styrax benzoin)*

Description: Onycha is a tree native to Indonesia. Styrax trees are evergreen trees with simple ovate leaves. It has flowers and fruit that droop and a vanilla-like aroma.

Verse: Exodus 30:34—"Then the LORD said to Moses, 'Take for yourself spices, stacte and onycha and galbanum, spices with pure frankincense; there shall be an equal part of each.'"

Scriptural Background: Onycha is only mentioned one time in scripture. It is listed as one of the ingredients in the Holy Incense (Exodus 30:34).

Historical Info: In ancient times, onycha was used in perfume, incense, and for wellness. It is thought that Phoenicians traded it in the 5th century BC. The larger trees were, and still are, used to make small, handmade decorative items. Today, it is used as an ornamental tree in west Africa, and as incense in the Middle East.

Modern Day Uses:
- Onycha is the thickest of all oils, so dilution is often necessary to thin the oil. Because of the thickness of this oil, it is not recommended for use in most diffusers.
- Onycha helps maintain the appearance of healthy, beautiful, youthful, radiant, luminous, and toned skin.
- Onycha is traditionally known for its comforting and soothing properties.
- It supports the respiratory, digestive, and nervous systems. It also helps maintain normal blood sugar levels, is soothing to the skin, and is great for massages.

There are two primary schools of thought regarding the source of onycha referenced in the Bible. The first and most widely held view is that 'onycha' is the strombus or wing-shell, which is a univalve snail that lives in the Indian Ocean. It is thought that the claw of the shell was burned in incense, creating a very strong odor like 'castoreum,' which is an odor used by perfumers to help create certain scents. Although snails are not considered kosher, this may still be a plausible option because some believe Jewish kosher laws pertain more to what one eats than with objects used for other purposes, including the sacred objects of the temple. For instance, there is a particular blue dye that has been re-introduced into Orthodox use for the tzizit that is made from a non-kosher snail and is approved for use as a dye. (The tzizit are the tassels that are on the four corners of the Jewish prayer shawl and everyday undergarment.) It is commanded in Numbers 15:38 that these be dyed in תְּכֵלֶת (tekeleth), which is a blue/violet dye that comes from a snail known as the Chilazon. This snail has been identified today as the *Murex trunculus*.[9]) The Chilazon dye was very precious and was literally worth its weight in gold. The reason for this particular dye color is because the color blue signifies God's glory (Exodus 24:9-10). It is quite possible that since onycha was likely burned as incense, it was not considered a sacrifice, but an offering.

> The second possible source of onycha is that of a flowering plant. In this case, the most probable options would be either *Cistus ladaniferus* or *Styrax benzoin*. Both of these are native to Asia and were widely used during that time.

Rose of Sharon

The oil is distilled from the branches of this vine like shrub.

Name: Rose of Sharon—plant type is highly debated see note on page 83*

Hebrew: חֲבַצֶּלֶת "rose, meadow saffron or crocus," שָׁרוֹן "Sharon" = crocus found on the plain of Sharon.

YL Equivalent: Rose of Sharon *(Cistus ladaniferus)*

Description: Rose of Sharon is a species of flowering plant that is native to the western Mediterranean region. Its common names include gum rockrose, laudanum, labdanum, common gum cistus, and brown-eyed rockrose. It is a shrub from the family of Cistaceae with broad, dark, evergreen leaves.

Verse: Song of Solomon 2:1—"I am the rose of Sharon, the lily of the valleys."

Scriptural Background: The words "rose" and "Sharon" are only mentioned next to each other once in scripture. In Song of Solomon 2:1, King Solomon's bride describes herself as the Rose of Sharon. This specific word for rose or crocus is mentioned twice in scripture (Song of Solomon 2:1 and Isaiah 35:1). The word used for Sharon is mentioned seven times in scripture (Joshua 12:8; I Chronicles 5:16; 27:29; Song of Solomon 2:1; Isaiah 33:9; 35:2; 65:10).

Historical Info: Rose of Sharon (cistus) was used in ancient times as incense and in perfumes. It was exported annually from Crete and Cypress. Its resin has been found in ancient tombs at the Acropolis in Athens. The Greeks also used it in cosmetics.

Modern Day Uses:
- Rose of Sharon is often used in perfumes.
- Cistus supports the body's natural response to irritation and injury. It has natural cleansing and purifying properties and may be applied neat or diluted.
- It supports the immune and cardiovascular systems and has been studied for its supportive effects on skin cell function.

It is still debated amongst botanists, scholars, and theologians as to what is actually referred to in this passage. So much time has distanced us from the source that many things are subject to speculation. In many cases, the Hebrew cannot be accurately translated, even by linguistic experts. The Hebrew word in Song of Solomon for rose is "chabatstseleth." The proper meaning of this word is "meadow saffron" or "crocus." The Hebrew word here for Sharon is indeed "Sharon," which is a plain on the Mediterranean Sea that still exists today. The Sharon Plain is located in Israel; its northern edge is the Gishon River, its southern edge is the Yargon River, its eastern edge is the Samaria hill country, and its western edge is the Mediterranean Sea. If it is the crocus, then the most probable crocus plant mentioned in this passage is either the Winter Crocus *(Crocus hyemalis)* or the *Crocus aleppicus*. If it is a meadow saffron, it is most probably the Steven's Meadow Saffron *(Colchicum stevenii)*, but this plant is very poisonous. Yet, others believe it could be some sort of lily or hibiscus.

Blends

The following blends include many of the Bible oils:

Exodus II™

Description: According to Exodus chapter 30, the Holy Anointing Oil contained essential oils of myrrh, cassia, cinnamon, and calamus in olive oil. Exodus II™, a Young Living blend created by Gary Young, contains all of these oils in addition to a few others. Notice that seven of the nine oils in this blend are Bible oils. No other blend has as many Bible oils as Exodus II™.

Ingredients: Exodus II™ contains olive oil, myrrh, cassia, cinnamon, calamus, hyssop, frankincense, northern black spruce, and vetiver.

Modern Day Uses: Exodus II™ has cleansing properties, particularly when applied to the feet. It also supports the respiratory system.

3 Wise Men™

Description: 3 Wise Men™ is a blend made with four Bible oils, one tree oil, and one carrier oil.

Ingredients: 3 Wise Men™ contains sweet almond oil, Royal Hawaiian Sandalwood, juniper, frankincense, spruce, and myrrh.

Modern Day Uses: It helps support a healthy pineal gland and promotes emotional well-being.

The Gift™

Description: The Gift™ is a blend of seven ancient oils that have been used for centuries in Arabia, several of which are Bible oils.

Ingredients: The Gift™ contains balsam needle oil, sacred frankincense, jasmine, northern lights black spruce, myrrh, vetiver, and cistus.

Modern Day Uses: It helps support the immune system.

Conclusion

As we have seen throughout this book, essential oils were very important in ancient times. They were important to the Israelites, Greeks, Romans, Egyptians, Assyrians, Babylonians, Indians, and many other people groups throughout Asia and Africa. Essential oils were used in a myriad of ways such as ingredients in food, incense, anointing, protection, wellness, etc. The oils we use today are not just a passing fad but are legitimately grounded in the history of the world. We can be assured that using essential oils today not only reflects a rich Biblical history, but can also enhance our journey towards wellness and spiritual abundance.

God mentions plants more than 600 times in His Word, and He has commanded us to learn about them, to study them, to protect them, and to use them. Plants are precious gifts from our Father in heaven and are meant to be used in a variety of ways. Even though some of the meanings and specific plants are beyond our grasp or full understanding, we must not let this deter us from using His gifts. The more we dig into scripture and learn about these plants, the more we will learn about God Himself, which opens an opportunity for us to grow deeper in our relationship with Him. So I encourage you to dig deep, open your heart, and see what God can and will do in your life.

All the Plants of the Bible

Acacia, also Shittah (Exodus 25:10)
Algum Tree also Sandalwood (2 Chronicles 2:8)
Almond (Genesis 43:11)
Aloe (Proverbs 7:17)
Apple (Proverbs 25:11)
Balm of Gilead (Jeremiah 8:22)
Bay also Laurel (Psalm 37:35)
Barley (Numbers 5:15)
Bdellium (Numbers 11:7)
Bean (Ezekiel 4:9)
Box Tree (Isaiah 41:19)
Bramble also Thorns (Isaiah 34:13)
Broom tree (bush) or Juniper tree (Psalm 120:4)
Cane, also Calamus (Isaiah 43:24)
Cassia (Psalm 45:8)
Cattail, Lotus plant, and Chaste tree (Job 40:21)
Cedar (I Kings 5:10)
Cinnamon (Exodus 30:23)
Citron Wood (Revelation 18:12)
Coriander (Numbers 11:7)
Cucumber (Numbers 11:5)
Cumin (Isaiah 28:27)
Cypress (Isaiah 44:14)
Dates (Song of Solomon 5:11)
Dill (Isaiah 28:27)

Ebony (Ezekiel 27:15)
Fig (Joel 1:7)
Fir Tree (Ezekiel 27:5)
Flax (Proverbs 31:13)
Flowers of the field (Isaiah 40:6)
Frankincense (*sacra* and *carteri*) (Leviticus 2:1-3)
Galbanum (Exodus 30:34)
Gall, Hemlock, or Wormwood (Amos 6:12)
Garlic (Numbers 11:5)
Gopher Wood (Genesis 6:14)
Gourd (II Kings 4:39)
Grape (Numbers 6:4)
Henna (Song of Solomon 1:14)
Hyssop also Caper (I Kings 4:33)
Leeks (Numbers 11:5)
Lentil (Ezekiel 4:9)
Lily (Matthew 6:28)
Mallow (Job 30:4)
Mandrake (Genesis 30:15)
Melon (Numbers 11:5)
Millet also Sorghum (Ezekiel 4:9)
Mint "Peppermint," Horsemint, and Garden Mint (Matthew 23:23)
Mulberry tree (Luke 17:6)
Mustard seed (Luke 17:6)
Myrrh (Exodus 30:23)
Myrtle (Nehemiah 8:15)
Nettles (Isaiah 34:13)
Oak (Joshua 24:26)
Olive (Judges 9:9)
Onion (Numbers 11:5)
Onycha, Storax, and Styrax (Exodus 30:34)
Palm Tree (Song of Solomon 7:8)
Pistachio, also Terebinth (Genesis 43:11)

Plane tree (Ezekiel 31:8)
Pomegranate (Song of Solomon 4:13)
Poplars (Isaiah 44:4)
Reed, also Papyrus (Exodus 2:3)
Rose of Sharon, Crocus, and Saffron (Song of Solomon 2:1)
Rue (Luke 11:42)
Rye (Isaiah 28:25)
Spelt (Ezekiel 4:9)
Spikenard, also Nard (Song of Solomon 4:13)
Sycamore Fig (Amos 7:14)
Tamarisk (Genesis 21:33)
Tares (Matthew 13:24-30)
Thistle, also Thorn (2 Chronicles 25:18)
Wheat (Genesis 30:14)
Willow (Psalm 137:2)

Explanation for some plants included in this list:

1. Gall, hemlock, and wormwood are listed together because they are all considered poisonous plants. Bible translators use these three names interchangeably, due to the fact that it is difficult to determine precisely which plant is being referenced in each passage.

2. Aloes and Sandalwood are listed separately in this list because it is most probable that they are two totally different trees. The Algum or Almug tree found in 2 Chronicles 2:8, 9:10-11 is most likely the Indian Sandalwood (*Santalum album*) found in India. Aloes, found in Proverbs 7:17, is a different tree altogether. It is most likely the Agarwood, also known as the Aloeswood, which is the resinous heartwood that comes from the endangered Aquilaria Malaccensis Tree.

Jewish Recipes Using Vitality™ Oils

Israeli Salad

Make 5-6 servings
This recipe uses many fresh foods that are common in Israel, along with spices and herbs used in ancient and modern times.

 6 cucumbers, diced
 4 plum or Roma tomatoes, diced
 3-4 green onions, sliced
 1 yellow bell pepper, diced
 3-4 cloves fresh garlic, minced
 1 cup chopped fresh parsley
 2 drops Peppermint Vitality™ essential oil
 1/2 cup good quality olive oil
 2 tablespoons fresh lemon juice
 1 tablespoon salt
 1 tablespoon black pepper

Place vegetables in a glass bowl; set aside. In a separate glass or metal bowl, add the rest of the ingredients and whisk to combine. Pour the dressing over the vegetable mixture and toss to coat. This dish tastes best when the flavors have had 20 minutes or more to combine.

Mohn Kihel Cookies

Historically a Ashkenazi Jewish recipe for the Sabbath, these wonderful, light cookies are perfect any day of the week with a cup of tea.

   3 cups organic all-purpose flour
   1/2 cup poppy seeds
   1 1/2 teaspoons baking powder
   1/4 teaspoon salt
   1 cup butter, softened
   2/3 cup organic cane sugar
   1 egg, separated
   2 tablespoon lemon zest
   2 drops Lemon Vitality™ essential oil
   1/4 cup lemon juice
   1 drop Coriander Vitality™ essential oil
   1/3 cup granulated sugar for decoration

Preheat oven to 350° F (175° C). Line baking sheets with parchment paper. Stir together flour, poppy seeds, baking powder, and salt; set aside.

In a medium glass or metal bowl, cream the butter and sugar together until light; beat in the egg yolk, lemon zest, both essential oils, and lemon juice. Fold in the flour mixture and mix well.

Divide dough in half, and roll each half out on a lightly floured surface until 1/8 to 1/4 inch thick. Cut with cookie cutters and place cookies on the prepared baking sheet. Brush tops of cookies with beaten egg white and sprinkle with white sugar.

Bake at 350° F for 12 to 15 minutes or until golden brown on the edges.

Tzimmes

This traditional vegetable side dish is popular amongst the Ashkenazi Jews. Sweet dishes are eaten to symbolize the wish for a good year ahead. It is also customary to eat foods that allude to blessings and prosperity. The Yiddish word for carrots is "meren," which means "to multiply."

- 1 large Spanish or sweet onion, diced
- 1/4 cup good quality olive oil
- 1 lb carrots, peeled and sliced in 1/2 inch rounds
- 10 prunes, diced
- 1 1/2 cup orange juice
- 1/2 cup honey
- 1 drop Cinnamon Vitality™ essential oil
- 1/2 teaspoon salt

Sauté the onion in olive oil over medium heat for about 20 minutes. Combine Cinnamon Vitality™ oil and honey in a small bowl. Add remaining ingredients to the saucepan with onion. Add honey and oil mixture, then simmer for 2 hours until tender. Serve warm.

Diffuser Blend Recipes

Ancient World
 3 drops Frankincense
 3 drops Myrrh
 3 drops Cedarwood
 3 drops Myrtle

Spice Market
 1 drop Cinnamon
 1 drop Cassia
 2 drops Coriander
 1 drop Peppermint

Plain of Sharon
 3 drops Laurus Nobilis
 3 drops Rose of Sharon
 1 drop Hyssop

Forest of Lebanon
 2 drops Cypress
 2 drops Cedarwood
 3 drops Juniper
 1 drop Royal Hawaiian Sandalwood

Land of Goshen
  2 drops Galbanum
  2 drops Spikenard
  2 drops Myrrh

Spirit of Jerusalem
  1 drop Hyssop
  2 drops Sacred Frankincense
  3 drops Myrtle
  1 drop Sacred Sandalwood

Endnotes

1. Steven D. Ehrlich, "Aromatherapy," *University of Maryland Medical Center* (August 2011): accessed January 9, 2017, http://umm.edu/health/medical/altmed/treatment/aromatherapy.

2. "Essential Oils in the Ancient World: Part II," *Young Living Blog* (April 15, 2015): accessed January 10, 2017, https://www.youngliving.com/blog/essential-oils-in-the-ancient-world-part-ii/.

3. The Editors of Encyclopedia Britannica, "Spice Trade," *Encyclopedia Britannica* (April 1, 2016): accessed January 7, 2017, https://www.britannica.com/topic/spice-trade.

4. W. E. Vine, Merrill F. Unger, and William White, *Vine's Complete Expository Dictionary of Old and New Testament Words: With Topical Index* (Nashville: T. Nelson, 1996), "Anoint," 28.

5. Isador Singer and Cyrus Adler, *The Jewish Encyclopedia: A Descriptive Record of the History, Religion, Literature, and Customs of the Jewish People from the Earliest times to the Present Day* (New York: Funk and Wagnalis, 1906), "Day of Atonement," http://www.jewishencyclopedia.com/articles/2093-atonement-day-of.

6   *Essential Oils: Desk Reference* (Orem, UT: Essential Science Publishing, 2015), 37-38.

7   "Mostar, Macao and Biblical Vestiges in Israel Are Among the 17 Cultural Sites Inscribed on UNESCO's World Heritage List," *UNESCO World Heritage Centre* (July 15, 2005): accessed January 10, 2017, http://whc.unesco.org/en/news/135/.

8   Jennie Cohen, "A Wise Man's Cure: Frankincense and Myrrh," *history.com. A&E Television Networks* (June 27, 2011): accessed January 7, 2017, http://www.history.com/news/a-wise-mans-cure-frankincense-and-myrrh.

9   "Tekhelet 101" *Ptil Tekhelet: accessed February 9, 2017,* http://tekhelet.com/tekhelet/introduction-to-tekhelet/.

Bibliography

Balfour, John Hutton. The Plants of the Bible 'Trees and Shrubs'. London: n.p., 1885.

Bromiley, G. W. *The International Standard Bible Encyclopedia*. Grand Rapids, MI: W.B. Eerdmans, 1979.

Cohen, Jennie. "A Wise Man's Cure: Frankincense and Myrrh." *History.com*. A&E Television Networks, 27 June 2011. Web. 07 Jan. 2017.

"Eastons Bible Dictionary Online." *Bible Study Tools*. N.p., n.d. Web. 07 Jan. 2017.

Ehrlich, Steven D. "Aromatherapy." *University of Maryland Medical Center*. N.p., 9 Aug. 2011. Web. 09 Jan. 2017.

Essential Oils: Desk Reference. Orem, UT: Essential Science Pub., 2015.

"Essential Oils in the Ancient World: Part II." *Young Living Blog*. N.p., 15 Apr. 2015. Web. 10 Jan. 2017.

Harmon, Catherine. "Pope Francis' Coat of Arms and Motto, Explained." Pope Francis' Coat of Arms and Motto, Explained | Catholic World Report—Global Church News and Views. Catholic World Report, 18 May 2013. Web. 20 Jan. 2017.

Hobbs, J. C. A Journey Through Albania and Other Provinces of Turkey in Europe and Asia to Constantinople. 3rd ed. Vol. 1. London: James Cawthorn, 1833.

Kowalchik, Claire, and William H. Hylton. *Rodale's Illustrated Encyclopedia of Herbs*. Emmaus, PA: Rodale, 1998.

Maisch, John M., ed. "Varieties." *American Journal of Pharmacy* 49.1 (1877).

"Mostar, Macao and Biblical Vestiges in Israel Are among the 17 Cultural Sites Inscribed on UNESCO's World Heritage List." *UNESCO World Heritage Centre*. N.p., 15 July 2005. Web. 10 Jan. 2017.

Musselman, Lytton John. "Plants of the Bible—ODU Plant Site." *Plants of the Bible—ODU Plant Site*. N.p., 27 Apr. 2007. Web. 07 Jan. 2017.

New American Standard Bible: The Lockman Foundation, 1995.

Penoel, Daniel, ed. Integrated Guide to Essential Oils and Aromatherapy. 2nd ed. N.p.: Sound Concepts, 2014.

Rhodes, Dianne Lee. "Cultural History of Three Traditional Hawaiian Sites (Chapter 5)." *National Parks Service*. U.S. Department of the Interior, 15 Nov. 2001. Web. 20 Jan. 2017.

"Royal Wedding Bouquet Returned to Abbey to Rest on the Grave of the Unknown Warrior." Westminster Abbey » Royal Wedding Bouquet Returned to Abbey to Rest on the Grave of the Unknown Warrior. N.p., n.d. Web. 09 Jan. 2017.

Schiller, Carol, and David Schiller. The Aromatherapy Encyclopedia: A Concise Guide to over 395 Plant Oils. Laguna Beach, CA: Basic Health Publications, 2012.

Singer, Isidor, and Cyrus Adler. The Jewish Encyclopedia: A Descriptive Record of the History, Religion, Literature, and Customs of the Jewish People from the Earliest times to the Present Day. New York: Funk and Wagnalis, 1906.

Stewart, David. *Healing Oils of the Bible*. Marble Hill, MO: Center for Aromatherapy Research & Education, 2002.

Strong, James. Strong's Exhaustive Concordance of the Bible. Peabody, MA: Hendrickson, 2007. www.biblehub.com

"Tekhelet 101." *Ptil Tekhelet*. Web. 09 Feb. 2017.

The Editors of Encyclopedia Britannica. "Spice Trade." *Encyclopedia Britannica*. Encyclopedia Britannica, Inc., 01 Apr. 2016. Web. 07 Jan. 2017.

Vine, W. E., Merrill F. Unger, and William White. Vine's Complete Expository Dictionary of Old and New Testament Words: With Topical Index. Nashville: T. Nelson, 1996.

Wilson, Roberta. A Complete Guide to Understanding and Using Aromatherapy for Vibrant Health and Beauty. Garden City Park: Avery Group, 1995.

About the Author

Joshua Graff, MAR

Joshua Graff has a Bachelors of Arts in Religion from East Texas Baptist University and a Masters of Arts in Religion from Liberty Baptist Theological Seminary. After serving as an associate pastor for over a decade and a brief stint in the US Navy Chaplaincy Candidate program, Joshua redirected his focus to help comfort others in their time of grief as a hospice chaplain for 5 years. Joshua married his wife Sharon in April 2016, and has joined her in building a business with Young Living Essential Oils. Josh's passions include helping people live a healthy lifestyle through exercise, clean eating, and natural health, while growing in their spiritual life through Biblical principles of leadership, servanthood, and discipleship. The author uses and highly recommends Young Living Essential Oils.

INDEX

Symbols
3 Wise Men™ 84

A
Aloes 33, 58, 59, 91
Anointing 23, 24, 25, 33, 35, 58, 84, 87
Aromatically 15, 43, 65
Assyrians 26, 40, 41, 47, 59, 71, 72, 87

B
Babylonians 26, 41, 47, 59, 71, 87
Bay tree 30, 31
Bible 2, 7, 8, 9, 11, 12, 13, 17, 23, 26, 27, 29, 30, 33, 35, 37, 39, 43, 47, 50, 52, 55, 58, 64, 67, 69, 70, 79, 84, 85, 89, 91, 101, 102, 103
Blends 12, 16, 84

C
Cassia 20, 27, 32, 33, 34, 58, 84, 89, 97
Cedar (all variations of the word) 29, 39, 40, 62, 69, 70, 71, 72, 73, 89
Cinnamon 20, 27, 32, 33, 34, 35, 36, 64, 84, 89, 95, 97

Condiment 27, 42, 43
Coriander 37, 38, 89, 94, 97
Creation 8, 11, 12, 29, 40, 70, 72
Cypress 39, 40, 41, 62, 70, 71, 72, 82, 89, 97

D
Dill 42, 43, 44, 54, 89
Distilled (all variations of the word) 29, 30, 32, 34, 37, 39, 42, 45, 49, 51, 53, 54, 57, 61, 64, 66, 69, 73, 77, 81

E
Egyptians 17, 26, 38, 40, 41, 43, 47, 50, 52, 59, 72, 73, 75, 87
Essential oils 7, 9, 12, 13, 15, 17, 18, 23, 26, 27, 33, 47, 55, 84, 87, 94
Exodus II™ 84

F
Frankincense 20, 45, 46, 47, 48, 50, 57, 59, 73, 77, 84, 85, 90, 97, 98, 100, 101

G
Galbanum 49, 50, 77, 90, 98
Gall 59, 90, 91
God 6, 8, 11, 12, 24, 25, 26, 29, 33, 35, 40, 41, 43, 46, 47, 52, 55, 62, 64, 65, 70, 71, 72, 79, 87, 109
Greeks 17, 23, 31, 33, 43, 53, 73, 75, 82, 87

H
Hebrew 8, 9, 25, 30, 32, 33, 34, 37, 39, 42, 45, 49, 51, 57, 61, 62, 64, 66, 67, 69, 74, 76, 77, 81, 83
Hyssop 51, 52, 53, 59, 69, 84, 90, 97, 98

I
Incense 20, 26, 45, 46, 47, 50, 59, 65, 67, 68, 75, 77, 78, 79, 82, 87
Indians (those from India) 17, 87
Internally 16
Israelites 25, 26, 27, 37, 40, 46, 47, 52, 62, 72, 87

J
Jesus (Yeshua) 23, 33, 43, 46, 47, 52, 55, 58, 59, 64, 65, 109, 110
Juniper 40, 62, 74, 75, 76, 84, 89, 97

K
King David 39, 70
King Hezekiah 13
King Solomon 35, 40, 41, 46, 52, 59, 65, 67, 70, 71, 72, 82

L
Laurus Nobilis 30, 97
Law (Torah) 43, 54, 62
Lord 11, 23, 40, 50, 62, 70, 72, 75, 109

M
Mint 27, 42, 54, 55, 90
Myrrh 20, 33, 35, 45, 47, 57, 58, 59, 60, 84, 85, 90, 97, 98, 100, 101
Myrtle 40, 61, 62, 63, 90, 97, 98

N
Native Americans 75

O
Oils 2, 7, 8, 9, 12, 13, 15, 16, 17, 18, 19, 23, 26, 27, 29, 30, 31, 33, 47, 50, 52, 55, 59, 60, 69, 78, 84, 85, 87, 93, 94, 99, 100, 101, 102, 103, 105
Onycha 50, 77, 78, 79, 80, 90

P

Peppermint 54, 55, 56, 90, 93, 97, 63

Perfume 17, 18, 24, 26, 33, 50, 58, 59, 64, 65, 78, 79, 82

Plain of Sharon 97

Plant 8, 9, 11, 12, 15, 17, 29, 31, 32, 37, 38, 39, 42, 43, 51, 52, 53, 54, 55, 57, 62, 64, 74, 76, 80, 81, 83, 89, 91, 101, 102, 110

Promise 40, 52, 62, 71, 72

Protection 24, 25, 87

R

Romans 17, 24, 33, 43, 53, 65, 75, 87, 110

Rose of Sharon 81, 82, 91, 97

S

Sandalwood 29, 66, 68, 84, 89, 91, 97, 98

Scripture / Scriptural 7, 8, 11, 12, 29, 31, 33, 35, 37, 39, 43, 46, 50, 52, 55, 58, 62, 64, 67, 70, 75, 77, 82, 87

Spikenard 64, 65, 91, 98

Supports / Supportive 63, 65, 68, 82

T

The Gift™ 85

Topically 15, 36, 38, 41, 44, 50, 65, 73, 76

Trade route 20, 27, 35, 47, 67

W

Wellness 7, 11, 12, 27, 78, 87

This book was written with the reality that the God of the Bible is the Creator and Savior of the world. If you do not know Jesus as Lord, I encourage you to read the verses below and seek to understand the God who created all these plants.

Isaiah 9:6
"For a child will be born to us, a son will be given to us; And the government will rest on His shoulders; And His name will be called Wonderful Counselor, Mighty God, Eternal Father, Prince of Peace."

Matthew 1:21-23
"She will bear a Son; and you shall call His name Jesus, for He will save His people from their sins." Now all this took place to fulfill what was spoken by the Lord through the prophet: "Behold, the virgin shall be with child and shall bear a Son, and they shall call His name Immanuel," which translated means, "God with us."

Romans 3:23
"For all have sinned and fall short of the glory of God."

Romans 6:23
"For the wages of sin is death, but the free gift of God is eternal life in Christ Jesus our Lord."

Ephesians 2:8-9
"For by grace you have been saved through faith; and that not of yourselves, it is the gift of God; not as a result of works, so that no one may boast."

The sketches in this book were done by Randal Sanders. For more about Randal and his graphic and artwork, please check out his Facebook page at https://www.facebook.com/randal.sanders.5 or contact him through email at randalsandersart@gmail.com.

To learn how Yeshua (Jesus) lived and taught, contact Judaic Studies Institute about classes. Visit their website at www.jsi-edu.org or email them at jsi.academics@jsi-edu.org.

To obtain additional copies and for more information on other books by Growing Healthy Homes, please visit our website at www.GrowingHealthyHomes.com

Nutrition 101: Choose Life!

Gentle Babies

Road to Royal: Roadmap to Success

Road to Wellness: Roadmap for a Lifestyle of Health

The ABC's of Building a Young Living Organization